INTRODUCTION TO IOT: CONCEPTS AND APPLICATIONS

Explore the Internet of Things, its architecture, protocols, and real-world applications

HARMEET SINGH

Preface

In an age where the digital and physical worlds converge at an unprecedented pace, the Internet of Things (IoT) stands out as a transformative force shaping our everyday lives. From smart homes that adjust to our preferences to cities that use data to enhance urban living, IoT is not just a technological trend; it is a new paradigm that promises to redefine how we interact with our environment and each other.

This book, *Introduction to IoT: Concepts and Applications*, is designed to be a comprehensive guide for those eager to understand the intricacies of this dynamic field. Whether you are a student venturing into the world of IoT, a professional seeking to enhance your skills, or a curious reader wanting to grasp the fundamentals, this book provides a structured approach to learning.

We begin with foundational concepts, breaking down the components that make up IoT ecosystems, including sensors, connectivity, and data analytics. Each chapter builds upon the last, offering clear explanations and practical examples that illustrate the relevance and application of IoT in various industries—from healthcare and agriculture to transportation and smart cities.

As we delve into the applications of IoT, we also address the challenges and ethical considerations that accompany this technological revolution. Privacy, security, and sustainability are critical topics that we must navigate as we embrace the vast potential of interconnected devices.

Through real-world case studies and engaging illustrations, we aim to inspire a deeper understanding of how IoT technologies can drive innovation, efficiency, and new business models. We encourage readers to think critically about the implications of IoT and envision the future of a connected world.

In crafting this book, we have drawn upon the expertise of industry professionals, academic scholars, and our own experiences in the field. Our hope is that this work serves not only as a valuable resource but also as a catalyst for further exploration and innovation in the realm of IoT.

We invite you to embark on this journey with us as we explore the concepts and applications of the Internet of Things. Together, let us discover how these interconnected systems can enhance our lives, improve industries, and foster a smarter, more connected world.

— **Harmeet Singh** —

Table of Contents

Chapter 1: Introduction to IoT 10

    1.1 What is the Internet of Things? 10

    1.2 Historical Background 10

    1.3 Key Components of IoT 11

    1.4 How IoT Works 12

    1.5 Benefits of IoT 12

    1.6 Challenges and Concerns 13

    1.7 Conclusion 14

Chapter 2: IoT Architecture 15

    2.1 Understanding IoT Architecture 15

    2.2 The Layers of IoT Architecture 15

        2.2.1 Perception Layer 15

        2.2.2 Network Layer 16

        2.2.3 Application Layer 17

    2.3 Additional Components of IoT Architecture 17

        2.3.1 Edge Computing 18

        2.3.2 Security Framework 18

        2.3.3 Interoperability Standards 18

    2.4 IoT Architecture Models 19

        2.4.1 Centralized Architecture 19

        2.4.2 Distributed Architecture 19

        2.4.3 Hybrid Architecture 19

    2.5 Conclusion 20

Chapter 3: Communication Protocols 21

    3.1 Introduction to Communication Protocols in IoT 21

    3.2 Types of Communication Protocols 21

        3.2.1 Wired Protocols 21

        3.2.2 Wireless Protocols 22

    3.3 Protocol Selection Criteria 24

        3.3.1 Data Rate Requirements 24

        3.3.2 Range 24

        3.3.3 Power Consumption 24

        3.3.4 Scalability 24

- 3.3.5 Security Features 25
- 3.4 Emerging Communication Protocols 25
  - 3.4.1 5G 25
  - 3.4.2 MQTT (Message Queuing Telemetry Transport) 25
  - 3.4.3 CoAP (Constrained Application Protocol) 26
- 3.5 Conclusion 26

Chapter 4: Sensors and Actuators 27
- 4.1 Introduction to Sensors and Actuators in IoT 27
- 4.2 Understanding Sensors 27
  - 4.2.1 Definition and Functionality 27
  - 4.2.2 Types of Sensors 27
  - 4.2.3 Working Principles of Sensors 28
- 4.3 Understanding Actuators 29
  - 4.3.1 Definition and Functionality 29
  - 4.3.2 Types of Actuators 29
  - 4.3.3 Working Principles of Actuators 30
- 4.4 Role of Sensors and Actuators in IoT Applications 30
  - 4.4.1 Smart Homes 30
  - 4.4.2 Industrial Automation 31
  - 4.4.3 Agriculture 31
  - 4.4.4 Healthcare 31
- 4.5 Challenges and Considerations 31
  - 4.5.1 Calibration and Accuracy 32
  - 4.5.2 Power Consumption 32
  - 4.5.3 Security 32
  - 4.5.4 Interoperability 32
- 4.6 Future Trends in Sensors and Actuators 32
  - 4.6.1 Integration with Artificial Intelligence 33
  - 4.6.2 Miniaturization 33
  - 4.6.3 Improved Connectivity 33
- 4.7 Conclusion 33

Chapter 5: Data Management in IoT 35
- 5.1 Introduction to Data Management in IoT 35
- 5.2 Data Acquisition 35

- 5.2.1 Understanding Data Acquisition 35
- 5.2.2 Methods of Data Acquisition 35
- 5.2.3 Challenges in Data Acquisition 36
- 5.3 Data Storage 36
  - 5.3.1 Overview of Data Storage Solutions 36
  - 5.3.2 Data Storage Models 37
  - 5.3.3 Challenges in Data Storage 37
- 5.4 Data Processing 38
  - 5.4.1 Overview of Data Processing Techniques 38
  - 5.4.2 Data Processing Approaches 38
  - 5.4.3 Challenges in Data Processing 38
- 5.5 Data Analysis 39
  - 5.5.1 Importance of Data Analysis 39
  - 5.5.2 Data Analysis Techniques 39
  - 5.5.3 Tools for Data Analysis 40
  - 5.5.4 Challenges in Data Analysis 40
- 5.6 Data Governance and Security 40
  - 5.6.1 Importance of Data Governance 40
  - 5.6.2 Key Components of Data Governance 41
  - 5.6.3 Data Security Challenges 41
- 5.7 Conclusion 42

Chapter 6: IoT Communication Protocols 43
- 6.1 Introduction to IoT Communication Protocols 43
- 6.2 Importance of Communication Protocols in IoT 43
  - 6.2.1 Interoperability 43
  - 6.2.2 Data Transmission Efficiency 44
  - 6.2.3 Scalability 44
  - 6.2.4 Security 44
- 6.3 Overview of IoT Communication Protocols 44
  - 6.3.1 Network Layer Protocols 45
  - 6.3.2 Application Layer Protocols 46
- 6.4 Choosing the Right Communication Protocol 48
  - 6.4.1 Application Requirements 48
  - 6.4.2 Device Capabilities 48

    6.4.3 Network Environment 48

    6.4.4 Security Considerations 48

  6.5 Challenges in IoT Communication Protocols 49

    6.5.1 Fragmentation of Protocols 49

    6.5.2 Scalability Issues 49

    6.5.3 Security Vulnerabilities 49

    6.5.4 Power Consumption 49

  6.6 Future Trends in IoT Communication Protocols 50

    6.6.1 Emergence of New Protocols 50

    6.6.2 Integration of AI and Machine Learning 50

    6.6.3 Standardization Efforts 50

    6.6.4 Enhanced Security Mechanisms 50

  6.7 Conclusion 51

Chapter 7: Security Challenges in IoT 52

  7.1 Introduction to IoT Security 52

  7.2 Understanding IoT Security Challenges 52

    7.2.1 Device Security Vulnerabilities 52

    7.2.2 Network Security Threats 53

    7.2.3 Data Privacy Concerns 53

  7.3 Implications of Security Challenges 54

    7.3.1 Economic Impact 54

    7.3.2 Safety Risks 54

    7.3.3 Erosion of Trust 55

  7.4 Strategies for Enhancing IoT Security 55

    7.4.1 Secure Device Design 55

    7.4.2 Network Security Measures 55

    7.4.3 User Education and Awareness 56

    7.4.4 Regulatory Compliance 56

  7.5 Future Trends in IoT Security 56

    7.5.1 Artificial Intelligence and Machine Learning 57

    7.5.2 Zero Trust Architecture 57

    7.5.3 Blockchain Technology 57

    7.5.4 Standardization of Security Protocols 57

  7.6 Conclusion 57

Chapter 8: Future Trends in IoT ... 59
8.1 Introduction ... 59
8.2 The Rise of Edge Computing ... 59
8.2.1 Definition and Importance ... 59
8.2.2 Benefits of Edge Computing in IoT ... 59
8.2.3 Use Cases ... 60
8.3 The Integration of Artificial Intelligence (AI) ... 60
8.3.1 AI and IoT Synergy ... 60
8.3.2 Benefits of AI in IoT ... 60
8.3.3 Use Cases ... 61
8.4 The Expansion of 5G Technology ... 61
8.4.1 Overview of 5G Technology ... 61
8.4.2 Impact of 5G on IoT ... 62
8.4.3 Use Cases ... 62
8.5 Sustainability and IoT ... 63
8.5.1 The Role of IoT in Promoting Sustainability ... 63
8.5.2 Benefits of IoT for Sustainability ... 63
8.5.3 Use Cases ... 63
8.6 Enhanced Security Measures ... 64
8.6.1 The Growing Focus on Security ... 64
8.6.2 Trends in IoT Security ... 64
8.6.3 Use Cases ... 64
8.7 Conclusion ... 65

Chapter 9: Real-World Applications of IoT ... 66
9.1 Introduction ... 66
9.2 Smart Homes ... 66
9.2.1 Overview ... 66
9.2.2 Key Use Cases ... 66
9.2.3 Benefits ... 67
9.3 Healthcare ... 67
9.3.1 Overview ... 67
9.3.2 Key Use Cases ... 67
9.3.3 Benefits ... 68
9.4 Industrial IoT (IIoT) ... 68

- 9.4.1 Overview 68
- 9.4.2 Key Use Cases 68
- 9.4.3 Benefits 69
- 9.5 Agriculture 69
  - 9.5.1 Overview 69
  - 9.5.2 Key Use Cases 69
  - 9.5.3 Benefits 70
- 9.6 Transportation and Logistics 70
  - 9.6.1 Overview 70
  - 9.6.2 Key Use Cases 70
  - 9.6.3 Benefits 71
- 9.7 Retail 71
  - 9.7.1 Overview 71
  - 9.7.2 Key Use Cases 71
  - 9.7.3 Benefits 72
- 9.8 Conclusion 72

Chapter 10: Challenges and Risks in IoT 73
- 10.1 Introduction 73
- 10.2 Security Vulnerabilities 73
  - 10.2.1 Overview 73
  - 10.2.2 Common Security Issues 73
  - 10.2.3 Potential Consequences 74
- 10.3 Data Privacy Concerns 74
  - 10.3.1 Overview 74
  - 10.3.2 Key Privacy Issues 75
  - 10.3.3 Potential Consequences 75
- 10.4 Interoperability Issues 75
  - 10.4.1 Overview 75
  - 10.4.2 Challenges to Interoperability 76
  - 10.4.3 Potential Consequences 76
- 10.5 Scalability Challenges 76
  - 10.5.1 Overview 77
  - 10.5.2 Key Scalability Issues 77
  - 10.5.3 Potential Consequences 77

10.6 Regulatory Compliance 78
   10.6.1 Overview 78
   10.6.2 Key Compliance Challenges 78
   10.6.3 Potential Consequences 78
10.7 Conclusion 79

Chapter 1: Introduction to IoT

1.1 What is the Internet of Things?

The Internet of Things (IoT) is a transformative technological paradigm that connects physical devices, vehicles, appliances, and various other objects to the internet, enabling them to collect and exchange data. This network of interconnected devices communicates seamlessly with each other, often without human intervention, creating an ecosystem where data flows freely and insights are generated in real-time. The concept of IoT integrates several fields, including electronics, data analytics, software development, and telecommunications, fundamentally changing how we interact with the world around us.

IoT can be defined simply as "the interconnection of uniquely identifiable embedded computing devices within the existing internet infrastructure." According to the International Telecommunication Union (ITU), IoT encompasses a wide range of technologies that enable the connection and communication of devices through various networks. This capability allows for enhanced monitoring, control, and automation of systems, making IoT an essential component of modern technological advancement.

1.2 Historical Background

The origins of IoT can be traced back to the early days of the internet and the development of embedded systems. The term "Internet of Things" was coined by Kevin Ashton in 1999 during a presentation at Procter & Gamble, where he discussed the potential of RFID (Radio Frequency Identification) technology to revolutionize supply chain management. Ashton's vision was a world where physical objects could be connected to the internet, providing real-time data and improving operational efficiencies.

The early 2000s saw significant advancements in sensor technology, wireless communication, and data analytics. These developments laid the foundation for the widespread adoption of IoT. By the mid-2010s, IoT had gained traction across various industries, including healthcare, agriculture, transportation, and smart home applications. The proliferation of smartphones and cloud computing further accelerated IoT's growth, enabling devices to collect, analyze, and share data seamlessly.

1.3 Key Components of IoT

The functionality of IoT relies on several key components:

1. **Devices and Sensors**: These are the physical objects embedded with sensors that collect data from their environment. Examples include temperature sensors, motion detectors, and cameras. Devices can range from simple household items to complex industrial machinery.
2. **Connectivity**: IoT devices must be connected to the internet to share data. This can be achieved through various communication protocols, such as Wi-Fi, Bluetooth, Zigbee, and cellular networks. The choice of connectivity depends on the application, range, and power consumption requirements.
3. **Data Processing**: Once data is collected, it needs to be processed to extract meaningful insights. This can occur on the device itself (edge computing) or in the cloud, where more extensive data analysis and storage capabilities are available.
4. **User Interface**: A user interface (UI) allows users to interact with the IoT system. This can be a mobile application, web dashboard, or other software interfaces that provide users with control and insights derived from the collected data.
5. **Analytics**: Advanced data analytics tools are essential for making sense of the vast amounts of data generated by IoT devices. Machine learning and artificial intelligence are often employed to identify patterns, predict outcomes, and enhance decision-making.

1.4 How IoT Works

The operation of an IoT system can be summarized in several steps:

1. **Data Collection**: Sensors and devices collect data from their surroundings. For instance, a temperature sensor in a smart thermostat monitors the room's temperature and sends the data to a cloud server.
2. **Data Transmission**: The collected data is transmitted over the internet to a central processing unit, which could be a cloud server or an edge device.
3. **Data Processing**: The data is processed to derive insights. This processing may involve filtering, aggregation, and analysis using algorithms or machine learning models.
4. **Action and Response**: Based on the analysis, actions can be taken automatically. For example, if a smoke detector senses smoke, it can send alerts to homeowners and emergency services.
5. **Feedback Loop**: The system learns from past data and outcomes to improve its responses and predictions. This feedback loop enhances the efficiency and effectiveness of IoT applications.

1.5 Benefits of IoT

The integration of IoT into everyday life offers numerous benefits:

1. **Increased Efficiency**: IoT automates routine tasks and processes, reducing the need for manual intervention and increasing overall operational efficiency. This is particularly evident in industrial settings, where IoT devices can optimize manufacturing processes.
2. **Enhanced Data Insights**: The ability to collect and analyze vast amounts of data enables organizations to make informed decisions based on real-time information, leading to improved business outcomes.

3. **Cost Savings**: By automating processes and optimizing resource usage, IoT can lead to significant cost savings for businesses and consumers alike. For example, smart meters in homes can help reduce energy consumption and costs.
4. **Improved Safety and Security**: IoT devices can enhance safety by monitoring environments and detecting anomalies. For instance, smart surveillance systems can alert homeowners to suspicious activity.
5. **Better Quality of Life**: In smart homes, IoT technologies improve convenience and comfort, enabling users to control devices remotely and automate daily routines.

1.6 Challenges and Concerns

While IoT offers numerous advantages, it also presents several challenges:

1. **Security Risks**: The interconnectivity of devices increases the risk of cyberattacks. Ensuring the security of IoT devices and networks is critical to prevent data breaches and unauthorized access.
2. **Data Privacy**: The collection of vast amounts of personal data raises concerns about privacy. Users must be informed about data collection practices and how their information is used.
3. **Interoperability**: With a wide variety of devices and communication protocols, ensuring interoperability between different IoT systems can be complex. Standards and regulations need to be established to facilitate seamless integration.
4. **Scalability**: As IoT networks expand, managing and scaling these systems can become challenging. Ensuring robust infrastructure is essential to support the growing number of connected devices.
5. **Environmental Impact**: The production and disposal of IoT devices can contribute to electronic waste. Sustainable practices must be adopted to minimize the environmental footprint of IoT technologies.

1.7 Conclusion

The Internet of Things is reshaping the landscape of technology, offering innovative solutions and opportunities across various sectors. As the number of connected devices continues to grow, so does the potential for enhancing efficiency, improving quality of life, and driving economic growth. However, addressing the challenges associated with IoT—particularly concerning security, privacy, and interoperability—will be crucial for its sustainable development. As we move forward into an increasingly interconnected world, understanding the fundamental concepts of IoT will be essential for harnessing its full potential and navigating the complexities it presents.

Chapter 2: IoT Architecture

2.1 Understanding IoT Architecture

The architecture of the Internet of Things (IoT) serves as the foundational blueprint that outlines how different components of an IoT system interact with one another. It consists of various layers, each responsible for specific functions, from data collection to processing and analysis. A well-designed IoT architecture is essential for ensuring seamless communication, efficient data handling, and the overall success of IoT applications.

IoT architecture can generally be categorized into three main layers: the perception layer, the network layer, and the application layer. Each layer plays a critical role in the operation of an IoT system, and understanding their functionalities is vital for anyone interested in leveraging IoT technologies.

2.2 The Layers of IoT Architecture

2.2.1 Perception Layer

The perception layer, also known as the physical layer, is the foundation of IoT architecture. It consists of the physical devices and sensors that collect data from the environment. This layer includes a variety of components, such as:

1. **Sensors**: Devices that detect and measure physical properties like temperature, humidity, motion, light, and pressure. Examples include temperature sensors in smart thermostats and motion detectors in security systems.
2. **Actuators**: Devices that take action based on commands received from the network. For example, a smart lock can be actuated to lock or unlock a door based on user input or a predefined schedule.

3. **RFID Tags**: Radio Frequency Identification (RFID) tags are used to automatically identify and track objects. They are commonly employed in supply chain management to monitor inventory.

The primary function of the perception layer is to gather data from the physical world and convert it into digital signals that can be processed by the subsequent layers. This data collection can be continuous or event-driven, depending on the specific application and the nature of the sensors deployed.

2.2.2 Network Layer

The network layer serves as the communication backbone of the IoT architecture. It is responsible for transmitting data collected from the perception layer to the processing units, such as cloud servers or edge devices. This layer includes:

1. **Communication Protocols**: The network layer employs various communication protocols to facilitate data transmission. Common protocols include Wi-Fi, Bluetooth, Zigbee, LoRaWAN, and cellular networks like 4G and 5G. The choice of protocol depends on factors such as range, data rate, power consumption, and network topology.
2. **Gateways**: IoT gateways act as intermediaries between the perception layer and the cloud or data center. They aggregate data from multiple devices, perform preliminary processing, and ensure secure transmission of data to the network. Gateways can also handle protocol translation, enabling devices with different communication standards to communicate with each other.
3. **Cloud Infrastructure**: The cloud serves as a central repository for data storage and processing. It offers scalable resources for handling large volumes of data generated by IoT devices. Cloud computing enables advanced analytics and machine learning algorithms to be applied to the data, generating valuable insights.

The network layer's role is crucial, as it ensures reliable, efficient, and secure communication between devices and the processing units. Given the diverse range of devices and communication standards in IoT, this layer must be designed with flexibility and scalability in mind.

2.2.3 Application Layer

The application layer is where the end-user interacts with the IoT system. It encompasses the software applications that provide the user interface and manage the functionality of IoT devices. Key components of the application layer include:

1. **User Interfaces**: Applications designed for various platforms (mobile, web, etc.) that allow users to monitor and control IoT devices. For example, a smartphone app for a smart thermostat enables users to adjust the temperature remotely.
2. **Data Analytics Tools**: Software that processes the data collected from IoT devices to generate insights and actionable information. These tools can employ machine learning algorithms to identify patterns, predict outcomes, and facilitate data-driven decision-making.
3. **Business Logic**: This refers to the rules and workflows governing the operation of IoT applications. It defines how data is processed, what actions are taken based on specific triggers, and how users interact with the system.

The application layer ultimately determines the value derived from the IoT system. A well-designed application can enhance user experience, improve operational efficiency, and provide meaningful insights that drive business decisions.

2.3 Additional Components of IoT Architecture

In addition to the three main layers of IoT architecture, several other components are critical for the successful implementation of IoT systems:

2.3.1 Edge Computing

Edge computing refers to the practice of processing data closer to the source of data generation, rather than relying solely on cloud-based processing. By performing data processing at the edge, such as on gateways or local servers, IoT systems can reduce latency, minimize bandwidth usage, and improve response times. Edge computing is especially beneficial for applications requiring real-time data processing, such as autonomous vehicles and industrial automation.

2.3.2 Security Framework

Security is a fundamental concern in IoT architecture. The increasing number of connected devices creates a larger attack surface for cyber threats. Therefore, implementing robust security measures at every layer of the IoT architecture is crucial. This includes:

1. **Data Encryption**: Encrypting data in transit and at rest to protect sensitive information from unauthorized access.
2. **Authentication and Access Control**: Implementing strong authentication mechanisms to verify the identity of users and devices. Role-based access control can help ensure that only authorized personnel can access specific data and functions.
3. **Regular Updates and Patching**: Keeping software and firmware up to date is vital for protecting IoT devices from known vulnerabilities.

2.3.3 Interoperability Standards

As IoT encompasses a diverse range of devices and technologies, ensuring interoperability is essential for seamless communication and data exchange. Establishing common standards and protocols can facilitate integration between different devices, applications, and platforms. Organizations such as the Open Connectivity Foundation (OCF) and the Internet Engineering Task Force (IETF) work toward creating interoperability standards for IoT systems.

2.4 IoT Architecture Models

Different IoT applications may require different architectural models. Here are a few common models:

2.4.1 Centralized Architecture

In a centralized architecture, all data processing and storage occur in a central cloud server. Devices send data to the cloud for processing, and users interact with the system through cloud-based applications. While this model offers scalability and ease of management, it may introduce latency and bandwidth challenges, particularly for real-time applications.

2.4.2 Distributed Architecture

Distributed architectures involve multiple processing nodes, such as edge devices and gateways, that perform data processing locally. This approach reduces latency and bandwidth requirements, making it suitable for applications that require real-time responsiveness. However, managing a distributed architecture can be more complex compared to a centralized approach.

2.4.3 Hybrid Architecture

A hybrid architecture combines elements of both centralized and distributed models. In this approach, some data is processed locally at the edge, while other data is sent to the cloud for more extensive analysis. This model provides a balance between real-time processing and advanced analytics, making it adaptable to various application needs.

2.5 Conclusion

Understanding the architecture of the Internet of Things is essential for harnessing its full potential. The perception, network, and application layers work together to facilitate seamless data collection, transmission, and analysis. As IoT continues to evolve, incorporating additional components such as edge computing and robust security measures will be critical for addressing the challenges posed by the increasing number of connected devices.

By carefully designing and implementing IoT architecture, organizations can create efficient, scalable, and secure IoT systems that deliver significant value across various industries. As we progress further into an era dominated by connectivity and automation, a solid understanding of IoT architecture will empower individuals and organizations to innovate and succeed in an increasingly digital world.

Chapter 3: Communication Protocols

3.1 Introduction to Communication Protocols in IoT

Communication protocols are essential for the Internet of Things (IoT) as they dictate how devices exchange information and communicate with one another. Given the diverse nature of IoT applications, ranging from industrial automation to smart homes, the choice of communication protocol can significantly affect the performance, scalability, and reliability of an IoT system.

Communication protocols can be categorized based on various factors, including the type of network (wired or wireless), range, power consumption, and data rate. This chapter will explore the most widely used communication protocols in IoT, discussing their features, advantages, disadvantages, and suitable applications.

3.2 Types of Communication Protocols

Communication protocols in IoT can be broadly divided into two categories: **wired** and **wireless** protocols. Each category serves different use cases and environments.

3.2.1 Wired Protocols

Wired protocols involve physical connections using cables or fiber optics. They are often used in industrial applications where reliability and speed are critical.

1. **Ethernet**: Ethernet is one of the most common wired protocols, providing high-speed data transfer over local area networks (LANs). It offers reliable communication with minimal latency and supports various data rates, from 10 Mbps to 100 Gbps. However, its reliance on physical cabling makes it less flexible than wireless alternatives. Ethernet is commonly used in environments where stable connections are crucial, such as factories and office buildings.

2. **Modbus**: Modbus is a communication protocol designed for industrial automation systems. It operates over serial lines (Modbus RTU) or TCP/IP networks (Modbus TCP). Modbus allows for communication between various devices, such as sensors, actuators, and controllers. It is simple, open-source, and widely adopted in SCADA (Supervisory Control and Data Acquisition) systems. However, its limited data rate and lack of advanced security features can be drawbacks in modern applications.

3.2.2 Wireless Protocols

Wireless protocols eliminate the need for physical connections, providing greater flexibility and mobility. They are widely used in consumer applications, smart homes, and industrial IoT.

1. **Wi-Fi**: Wi-Fi is a widely used wireless protocol that enables high-speed internet access over short distances. With data rates reaching up to several gigabits per second, Wi-Fi is suitable for applications requiring substantial bandwidth, such as video streaming and large file transfers. However, Wi-Fi can consume considerable power, making it less suitable for battery-operated devices. Additionally, its range is limited to approximately 100 meters indoors, which can be a challenge for larger deployments.

2. **Bluetooth**: Bluetooth is a short-range wireless protocol primarily used for connecting personal devices. It operates in the 2.4 GHz frequency band and is designed for low power consumption, making

it ideal for wearable devices and smart home applications. Bluetooth Low Energy (BLE), a variant of Bluetooth, is specifically optimized for low-power IoT devices, allowing them to operate for extended periods on small batteries. However, Bluetooth's limited range (typically around 10 to 100 meters) can restrict its use in certain applications.

3. **Zigbee**: Zigbee is a wireless protocol designed for low-power, low-data-rate applications. It operates in the 2.4 GHz frequency band and is commonly used in home automation, smart lighting, and sensor networks. Zigbee supports mesh networking, allowing devices to communicate with each other over long distances by relaying messages. This feature enhances reliability and coverage. However, Zigbee's data rate (up to 250 Kbps) is relatively low compared to other protocols, limiting its use in bandwidth-intensive applications.

4. **LoRaWAN (Long Range Wide Area Network)**: LoRaWAN is a wireless communication protocol designed for long-range, low-power IoT applications. It is particularly well-suited for scenarios requiring extensive coverage, such as smart agriculture, environmental monitoring, and smart city applications. LoRaWAN operates in sub-GHz frequencies and can transmit data over distances of up to 15 kilometers in rural areas. However, its low data rate (typically under 50 Kbps) makes it unsuitable for applications requiring real-time data transmission.

5. **NB-IoT (Narrowband IoT)**: NB-IoT is a cellular communication protocol specifically designed for IoT applications. It operates in licensed frequency bands and provides deep coverage, low power consumption, and a long battery life for connected devices. NB-IoT is ideal for applications such as smart meters, asset tracking, and industrial automation. However, it requires a cellular network infrastructure, which may not be available in all regions.

3.3 Protocol Selection Criteria

Selecting the appropriate communication protocol for an IoT application is a critical decision that depends on several factors:

3.3.1 Data Rate Requirements

Different applications have varying data rate requirements. High-bandwidth applications, such as video streaming or large data transfers, may necessitate protocols like Wi-Fi, while low-bandwidth applications, such as sensor data transmission, can effectively utilize protocols like Zigbee or LoRaWAN.

3.3.2 Range

The required range of communication is another essential consideration. For example, smart home devices may only need to communicate within a short distance, making Bluetooth or Zigbee suitable. In contrast, applications like agricultural monitoring may require long-range connectivity, making LoRaWAN a better choice.

3.3.3 Power Consumption

Power consumption is particularly critical for battery-operated devices. Protocols like Zigbee and BLE are designed for low power consumption, enabling devices to operate for extended periods on small batteries. In contrast, Wi-Fi, while powerful, may drain battery life quickly.

3.3.4 Scalability

As IoT networks grow, scalability becomes a key consideration. Protocols supporting mesh networking, like Zigbee, allow for easy expansion of networks without significant infrastructure changes. In contrast, centralized protocols may face challenges when scaling to accommodate numerous devices.

3.3.5 Security Features

Security is a paramount concern in IoT, given the increasing number of cyber threats. Protocols with built-in security features, such as encryption and authentication, are essential for protecting sensitive data and maintaining user privacy. Choosing a protocol with robust security measures should be a priority when designing IoT systems.

3.4 Emerging Communication Protocols

As IoT continues to evolve, several emerging communication protocols are gaining traction:

3.4.1 5G

5G technology promises high-speed, low-latency connectivity, making it a game-changer for IoT applications. With the ability to support a massive number of connected devices simultaneously, 5G is expected to revolutionize industries such as autonomous vehicles, smart cities, and remote healthcare. Its advanced capabilities will enable real-time data transmission and processing, opening up new possibilities for IoT innovation.

3.4.2 MQTT (Message Queuing Telemetry Transport)

MQTT is a lightweight messaging protocol designed for low-bandwidth and high-latency networks. It operates on a publish-subscribe model, allowing devices to communicate efficiently. MQTT is particularly well-suited for applications requiring real-time updates, such as monitoring and control systems. Its low overhead and simplicity make it a popular choice for IoT applications.

3.4.3 CoAP (Constrained Application Protocol)

CoAP is a specialized protocol designed for constrained devices and networks. It operates over UDP (User Datagram Protocol), making it lightweight and suitable for low-power and lossy networks. CoAP is ideal for applications where resources are limited, such as in smart homes and industrial IoT. Its ability to operate in constrained environments makes it a valuable addition to the IoT protocol landscape.

3.5 Conclusion

Communication protocols are the backbone of the Internet of Things, facilitating data exchange and enabling seamless interactions between devices. The choice of protocol can significantly impact the performance, scalability, and security of IoT systems. Understanding the characteristics, advantages, and limitations of various communication protocols is crucial for designing effective IoT applications.

As IoT continues to evolve, the emergence of new communication protocols will shape the future of connectivity. By carefully selecting the appropriate protocols based on application requirements, organizations can create efficient, reliable, and secure IoT solutions that unlock the full potential of this transformative technology.

Chapter 4: Sensors and Actuators

4.1 Introduction to Sensors and Actuators in IoT

Sensors and actuators are fundamental components of the Internet of Things (IoT) architecture, serving as the interface between the physical world and the digital realm. Sensors collect data from the environment, while actuators translate that data into physical actions. Together, these devices enable IoT systems to monitor, control, and automate processes across various applications, from smart homes to industrial automation.

In this chapter, we will explore the different types of sensors and actuators used in IoT, their working principles, applications, and the significance of their role in the broader context of IoT solutions.

4.2 Understanding Sensors

4.2.1 Definition and Functionality

Sensors are devices that detect changes in physical or environmental conditions and convert this information into signals that can be read and interpreted by computers or other electronic devices. Sensors play a crucial role in gathering real-time data, which is essential for making informed decisions in IoT applications.

4.2.2 Types of Sensors

Sensors can be classified based on various criteria, including the type of data they measure, the technology used, and their application domains. Below are some common types of sensors used in IoT:

1. **Temperature Sensors**: These sensors measure ambient temperature and can be used in various applications, including HVAC systems,

smart thermostats, and environmental monitoring. Common types include thermocouples, thermistors, and infrared sensors.

2. **Humidity Sensors**: Humidity sensors, or hygrometers, measure the moisture content in the air. They are widely used in HVAC systems, weather stations, and smart agriculture to ensure optimal environmental conditions.

3. **Light Sensors**: These sensors detect light intensity and are commonly used in smart lighting systems. They can automatically adjust the brightness of lights based on ambient light conditions, improving energy efficiency.

4. **Proximity Sensors**: Proximity sensors detect the presence of nearby objects without physical contact. They are often used in smartphones for screen activation, in security systems, and in industrial automation for detecting the position of objects on a conveyor belt.

5. **Pressure Sensors**: These sensors measure pressure levels in gases or liquids. They are used in various applications, including weather forecasting, industrial automation, and automotive systems.

6. **Motion Sensors**: Motion sensors detect movement and are widely used in security systems, smart home devices, and fitness trackers. They can trigger alarms or notifications when unexpected motion is detected.

4.2.3 Working Principles of Sensors

The working principles of sensors vary based on their type and application. Most sensors operate on the following fundamental steps:

1. **Sensing**: The sensor detects a specific physical phenomenon (e.g., temperature, light, motion) using various technologies, such as resistive, capacitive, or inductive methods.

2. **Signal Conversion**: The detected phenomenon is converted into an electrical signal. For example, a temperature sensor may produce a voltage output proportional to the measured temperature.

3. **Data Transmission**: The converted signal is transmitted to a microcontroller, gateway, or cloud server for further processing and analysis.
4. **Data Interpretation**: The received data is interpreted, analyzed, and acted upon based on predefined criteria or user input.

4.3 Understanding Actuators

4.3.1 Definition and Functionality

Actuators are devices that convert electrical signals into physical actions or movements. They are essential for controlling systems in response to the data collected by sensors. Actuators enable automation in various applications, allowing for real-time responses to changing conditions.

4.3.2 Types of Actuators

Actuators can be classified into different categories based on their operating principles and applications:

1. **Electric Actuators**: These actuators use electrical energy to create motion. They are commonly used in applications such as robotic arms, automated doors, and conveyor systems. Electric actuators can be further divided into linear actuators (which produce linear motion) and rotary actuators (which produce rotational motion).
2. **Pneumatic Actuators**: Pneumatic actuators use compressed air to generate motion. They are widely used in industrial automation, material handling, and robotics. Pneumatic systems are known for their speed and force, making them suitable for applications requiring rapid movement.
3. **Hydraulic Actuators**: These actuators utilize hydraulic fluid to produce motion. Hydraulic actuators are often used in heavy

machinery, construction equipment, and manufacturing processes due to their ability to generate significant force.

4. **Thermal Actuators**: Thermal actuators operate based on temperature changes. They are commonly used in thermostats and temperature control systems. As the temperature changes, the actuator responds by opening or closing valves to regulate heating or cooling.

4.3.3 Working Principles of Actuators

The operation of actuators can be summarized in the following steps:

1. **Signal Reception**: Actuators receive an electrical signal or command from a controller, microcontroller, or sensor.
2. **Energy Conversion**: The actuator converts the received electrical signal into mechanical energy. This can involve movement of a motor, the application of hydraulic or pneumatic pressure, or thermal expansion.
3. **Movement**: The actuator performs a specific action, such as opening a valve, moving a robotic arm, or adjusting the position of a device.
4. **Feedback**: In many cases, actuators are equipped with feedback mechanisms (such as position sensors) to provide information on their status and ensure accurate control.

4.4 Role of Sensors and Actuators in IoT Applications

4.4.1 Smart Homes

In smart homes, sensors and actuators work together to create a connected environment. For example, motion sensors can detect the presence of occupants in a room, while actuators control lighting and HVAC systems to optimize comfort and energy efficiency. Home automation systems can be programmed to adjust settings based on user preferences or environmental conditions, enhancing convenience and reducing energy consumption.

4.4.2 Industrial Automation

Sensors and actuators play a crucial role in industrial automation by enabling real-time monitoring and control of processes. For instance, temperature and pressure sensors can monitor equipment conditions, while actuators control valves and motors to maintain optimal operating parameters. This integration enhances operational efficiency, reduces downtime, and improves safety in manufacturing environments.

4.4.3 Agriculture

In smart agriculture, sensors are used to monitor soil moisture, temperature, and humidity levels. Actuators can then control irrigation systems, ensuring crops receive the right amount of water. This approach not only optimizes resource usage but also contributes to sustainable farming practices by reducing water waste and improving crop yields.

4.4.4 Healthcare

In the healthcare sector, sensors and actuators are integrated into medical devices for remote monitoring and patient care. Wearable sensors can track vital signs, while actuators control drug delivery systems or automated diagnostic devices. This integration allows healthcare professionals to monitor patients in real time, improving the quality of care and enabling timely interventions.

4.5 Challenges and Considerations

While sensors and actuators are essential components of IoT systems, several challenges must be addressed:

4.5.1 Calibration and Accuracy

Sensors must be calibrated regularly to ensure accurate measurements. Environmental factors, aging, and wear can affect sensor performance. Implementing routine calibration processes and using high-quality sensors can help mitigate these issues.

4.5.2 Power Consumption

Many IoT applications rely on battery-powered sensors and actuators. Power consumption is a critical consideration, particularly for devices that require long-term operation. Utilizing low-power components and optimizing communication protocols can extend battery life.

4.5.3 Security

Security is a significant concern in IoT systems. Sensors and actuators can be vulnerable to cyberattacks, potentially compromising data integrity and system functionality. Implementing robust security measures, such as encryption, authentication, and secure communication protocols, is essential to protect against threats.

4.5.4 Interoperability

The diverse range of sensors and actuators available in the market can lead to compatibility issues. Ensuring interoperability between different devices and protocols is crucial for creating cohesive IoT systems. Adopting industry standards and protocols can help facilitate this integration.

4.6 Future Trends in Sensors and Actuators

As IoT technology continues to advance, several trends are shaping the future of sensors and actuators:

4.6.1 Integration with Artificial Intelligence

The integration of artificial intelligence (AI) with sensors and actuators is poised to revolutionize IoT applications. AI algorithms can analyze data collected by sensors, enabling predictive maintenance, automated decision-making, and enhanced responsiveness. This combination enhances the overall efficiency and effectiveness of IoT systems.

4.6.2 Miniaturization

Advancements in technology are leading to the miniaturization of sensors and actuators. Smaller devices can be deployed in a wider range of applications, including wearables and smart home products. This trend enables greater flexibility in design and implementation.

4.6.3 Improved Connectivity

The evolution of communication protocols, such as 5G and LPWAN (Low Power Wide Area Network), is improving connectivity for sensors and actuators. Enhanced connectivity enables real-time data transmission, remote control, and greater scalability for IoT applications.

4.7 Conclusion

Sensors and actuators are integral components of the Internet of Things, enabling the collection of data and the execution of physical actions based on that data. Their ability to interface with the physical world allows IoT systems to monitor, control, and automate processes across various applications, enhancing efficiency, safety, and convenience.

As technology continues to evolve, the future of sensors and actuators holds exciting possibilities, including integration with AI, miniaturization, and improved connectivity. Addressing the challenges associated with calibration, power consumption, security, and interoperability will be critical for maximizing the potential of sensors and actuators in the rapidly growing IoT landscape.

Chapter 5: Data Management in IoT

5.1 Introduction to Data Management in IoT

The Internet of Things (IoT) generates vast amounts of data from various connected devices and sensors. Effective data management is crucial for harnessing this information to gain insights, make informed decisions, and optimize operations. As IoT continues to grow, the challenge of managing, storing, and analyzing data becomes increasingly complex. This chapter explores the fundamental aspects of data management in IoT, including data acquisition, storage, processing, and analysis, as well as the importance of data governance and security.

5.2 Data Acquisition

5.2.1 Understanding Data Acquisition

Data acquisition is the first step in the data management process, involving the collection of data from various sources, such as sensors, actuators, and devices. In IoT, data acquisition can occur in real-time or at scheduled intervals, depending on the application and requirements.

5.2.2 Methods of Data Acquisition

1. **Direct Measurement**: This method involves collecting data directly from sensors or devices. For example, a temperature sensor measures ambient temperature and transmits the data to a central system for analysis.
2. **Polling**: In polling, a central controller periodically requests data from connected devices. This approach is common in applications where

data changes infrequently, ensuring that the system does not overwhelm the network with constant data transmission.

3. **Event-Driven**: Event-driven data acquisition occurs when specific conditions trigger data transmission. For instance, a motion sensor may send an alert only when it detects movement, reducing unnecessary data traffic.

5.2.3 Challenges in Data Acquisition

1. **Data Quality**: Ensuring the accuracy and reliability of data collected from sensors is crucial. Calibration and maintenance of sensors are essential to mitigate data quality issues.
2. **Data Volume**: The sheer volume of data generated by IoT devices can overwhelm storage and processing capabilities. Efficient data acquisition strategies must be implemented to manage this influx of information.
3. **Latency**: In real-time applications, minimizing latency in data transmission is critical. Factors such as network congestion and device response times can impact overall system performance.

5.3 Data Storage

5.3.1 Overview of Data Storage Solutions

Once data is acquired, it must be stored efficiently for further processing and analysis. Data storage solutions can be broadly classified into two categories: **on-premises** and **cloud-based**.

1. **On-Premises Storage**: This involves storing data on local servers or devices. On-premises storage offers greater control over data security and access but may require significant investment in hardware and maintenance.

2. **Cloud-Based Storage**: Cloud storage solutions provide scalable and flexible options for storing vast amounts of data. They allow organizations to pay for only the storage they use and facilitate easy access and sharing of data across multiple devices.

5.3.2 Data Storage Models

1. **Structured Data Storage**: Structured data is organized in predefined formats, such as tables in relational databases. This model is suitable for applications requiring complex queries and relationships between data points.
2. **Unstructured Data Storage**: Unstructured data, such as images, videos, and text, lacks a predefined format. This type of data can be stored in NoSQL databases, which offer flexibility and scalability for handling diverse data types.
3. **Time-Series Data Storage**: IoT generates a significant amount of time-series data, which represents data points indexed in time order. Time-series databases are optimized for handling and analyzing such data, making them ideal for IoT applications that monitor metrics over time.

5.3.3 Challenges in Data Storage

1. **Scalability**: As the number of connected devices increases, the storage system must scale accordingly. Solutions should be designed to accommodate future growth without significant disruptions.
2. **Data Redundancy**: To ensure data integrity, redundancy measures must be implemented. However, excessive redundancy can lead to increased storage costs and complexity.
3. **Data Retrieval**: Efficient data retrieval mechanisms are essential for quickly accessing the information needed for analysis. Poor retrieval performance can hinder decision-making processes.

5.4 Data Processing

5.4.1 Overview of Data Processing Techniques

Data processing involves transforming raw data into meaningful information that can be analyzed and utilized. This step is critical for extracting insights and driving decision-making in IoT applications.

5.4.2 Data Processing Approaches

1. **Edge Computing**: Edge computing processes data closer to the source of generation (i.e., IoT devices) rather than sending it to a centralized cloud server. This approach reduces latency, minimizes bandwidth usage, and enhances real-time decision-making capabilities. Edge computing is particularly beneficial for applications requiring immediate responses, such as autonomous vehicles and industrial automation.

2. **Cloud Computing**: In cloud computing, data is sent to a remote server for processing. This approach leverages the scalability and computational power of cloud resources to handle large volumes of data. Cloud-based processing is suitable for applications involving complex analytics and machine learning.

3. **Hybrid Processing**: Hybrid processing combines both edge and cloud computing, allowing for a balance between real-time processing at the edge and more extensive analysis in the cloud. This approach enables organizations to optimize resources based on specific application requirements.

5.4.3 Challenges in Data Processing

1. **Latency**: In applications requiring real-time processing, minimizing latency is critical. Inefficient processing can lead to delays in decision-making and negatively impact system performance.

2. **Data Overload**: The high volume of data generated by IoT devices can overwhelm processing capabilities. Organizations must implement strategies to filter and prioritize data to ensure efficient processing.
3. **Complexity**: Data processing can involve complex algorithms and models, requiring specialized expertise and resources. Organizations must invest in the right tools and skills to effectively manage data processing tasks.

5.5 Data Analysis

5.5.1 Importance of Data Analysis

Data analysis is the process of examining and interpreting data to derive meaningful insights and inform decision-making. In the context of IoT, data analysis enables organizations to optimize operations, improve efficiency, and enhance user experiences.

5.5.2 Data Analysis Techniques

1. **Descriptive Analytics**: Descriptive analytics involves summarizing historical data to identify trends and patterns. This technique is useful for understanding past behaviors and performance metrics.
2. **Predictive Analytics**: Predictive analytics uses statistical algorithms and machine learning techniques to forecast future events based on historical data. This approach enables organizations to anticipate issues and make proactive decisions.
3. **Prescriptive Analytics**: Prescriptive analytics provides recommendations for actions based on data analysis. This technique helps organizations optimize processes and resource allocation by suggesting the best course of action.

5.5.3 Tools for Data Analysis

Various tools and platforms are available for data analysis in IoT, including:

1. **Business Intelligence (BI) Tools**: BI tools, such as Tableau and Power BI, enable organizations to visualize data and generate reports, making it easier to interpret and communicate insights.
2. **Machine Learning Platforms**: Machine learning platforms, such as TensorFlow and Scikit-learn, allow organizations to develop and deploy predictive models using IoT data.
3. **Data Analytics Platforms**: Platforms like Apache Spark and Hadoop provide scalable frameworks for processing and analyzing large datasets.

5.5.4 Challenges in Data Analysis

1. **Data Privacy**: Analyzing data from IoT devices can raise privacy concerns, particularly when sensitive information is involved. Organizations must implement data anonymization and protection measures to address these concerns.
2. **Data Silos**: Data from different sources may be stored in silos, making it challenging to perform comprehensive analysis. Integrating data across platforms and systems is essential for gaining holistic insights.
3. **Skill Gaps**: Analyzing complex IoT data requires specialized skills in data science and analytics. Organizations must invest in training and development to build the necessary expertise.

5.6 Data Governance and Security

5.6.1 Importance of Data Governance

Data governance refers to the policies, processes, and standards that ensure the quality, integrity, and security of data. Effective data governance is essential for managing the vast amounts of data generated by IoT devices and ensuring compliance with regulations.

5.6.2 Key Components of Data Governance

1. **Data Quality**: Ensuring data accuracy, consistency, and completeness is critical for effective decision-making. Organizations must implement data quality management processes to monitor and improve data quality.
2. **Data Ownership**: Clearly defining data ownership and responsibilities is essential for accountability and effective management. Organizations should establish roles for data stewards and custodians to oversee data governance efforts.
3. **Compliance**: Compliance with data protection regulations, such as GDPR and HIPAA, is crucial for organizations handling sensitive data. Implementing data governance frameworks helps ensure adherence to legal requirements.

5.6.3 Data Security Challenges

1. **Cybersecurity Threats**: IoT devices are often vulnerable to cyberattacks, making data security a significant concern. Organizations must implement robust security measures to protect data from unauthorized access and breaches.
2. **Data Encryption**: Encrypting data both in transit and at rest is essential for safeguarding sensitive information. Implementing encryption protocols helps protect data integrity and confidentiality.
3. **Access Control**: Implementing access control measures ensures that only authorized personnel can access sensitive data. Role-based access control (RBAC) can help manage user permissions effectively.

5.7 Conclusion

Data management is a critical aspect of the Internet of Things, encompassing data acquisition, storage, processing, analysis, and governance. As IoT continues to evolve, organizations must develop robust data management strategies to effectively harness the vast amounts of data generated by connected devices.

By implementing efficient data acquisition methods, choosing appropriate storage solutions, optimizing processing techniques, and ensuring strong data governance and security measures, organizations can unlock the full potential of IoT data. The insights gained from effective data management can drive innovation, improve operational efficiency, and enhance overall decision-making in an increasingly connected world.

Chapter 6: IoT Communication Protocols

6.1 Introduction to IoT Communication Protocols

In the Internet of Things (IoT), communication protocols serve as the foundational layer that enables devices to connect, communicate, and share data with one another. These protocols dictate how data is transmitted between devices, ensuring reliability, efficiency, and security in data exchange. As IoT continues to expand, the importance of robust communication protocols becomes increasingly evident, allowing for seamless interaction across diverse devices and applications.

This chapter will explore the various communication protocols used in IoT, their characteristics, applications, and challenges. Understanding these protocols is essential for designing effective IoT systems that can operate efficiently in a connected environment.

6.2 Importance of Communication Protocols in IoT

Communication protocols play a crucial role in the functioning of IoT systems for several reasons:

6.2.1 Interoperability

With a wide variety of devices from different manufacturers, interoperability is essential in IoT. Communication protocols ensure that devices can communicate effectively, regardless of their underlying technology. This interoperability allows for the integration of devices into a cohesive system, enabling them to work together seamlessly.

6.2.2 Data Transmission Efficiency

Different applications have varying data transmission requirements. Some applications may require real-time data exchange, while others can tolerate delays. Communication protocols are designed to optimize data transmission based on these requirements, balancing factors such as bandwidth, latency, and power consumption.

6.2.3 Scalability

As IoT networks grow, communication protocols must support an increasing number of devices and connections. Scalability is vital to ensure that the network can handle the growing volume of data and maintain performance without degradation.

6.2.4 Security

IoT devices often operate in environments where data security is paramount. Communication protocols include security features that protect data during transmission, ensuring confidentiality, integrity, and authentication. Robust security mechanisms help prevent unauthorized access and data breaches.

6.3 Overview of IoT Communication Protocols

IoT communication protocols can be broadly classified into two categories: **network layer protocols** and **application layer protocols**. Each category serves distinct purposes and offers various functionalities.

6.3.1 Network Layer Protocols

Network layer protocols are responsible for managing data transmission between devices over a network. Some of the prominent network layer protocols in IoT include:

6.3.1.1 MQTT (Message Queuing Telemetry Transport)

1. **Overview**: MQTT is a lightweight messaging protocol designed for low-bandwidth, high-latency environments. It follows a publish/subscribe model, allowing devices to send and receive messages efficiently.
2. **Characteristics**:
    1. Low overhead, making it ideal for constrained devices.
    2. Supports Quality of Service (QoS) levels to ensure message delivery reliability.
    3. Uses a central broker to manage message distribution.
3. **Applications**: MQTT is commonly used in home automation, industrial monitoring, and remote sensing applications.

6.3.1.2 CoAP (Constrained Application Protocol)

1. **Overview**: CoAP is designed for resource-constrained devices and low-power networks. It is based on the REST architecture and uses UDP (User Datagram Protocol) for communication.
2. **Characteristics**:
    1. Supports low overhead and efficient communication.
    2. Enables multicast communication, allowing messages to be sent to multiple devices simultaneously.
    3. Designed for use in constrained environments with limited processing power and memory.
3. **Applications**: CoAP is widely used in smart home devices, sensor networks, and industrial automation.

6.3.1.3 AMQP (Advanced Message Queuing Protocol)

1. **Overview**: AMQP is a robust messaging protocol that provides a reliable and secure messaging framework. It supports a variety of messaging patterns and is designed for enterprise-level applications.
2. **Characteristics**:
    1. Offers features such as message routing, queuing, and transactions.
    2. Provides strong security mechanisms, including encryption and authentication.
    3. Supports both point-to-point and publish/subscribe messaging patterns.
3. **Applications**: AMQP is used in financial services, supply chain management, and enterprise integration.

6.3.2 Application Layer Protocols

Application layer protocols define the structure and format of the messages exchanged between devices. Some of the widely used application layer protocols in IoT include:

6.3.2.1 HTTP/HTTPS (Hypertext Transfer Protocol/Secure)

1. **Overview**: HTTP is a widely used protocol for transmitting hypertext over the web. HTTPS is the secure version of HTTP, providing encryption and security for data transmission.
2. **Characteristics**:
    1. Supports RESTful API design, allowing for stateless communication.
    2. Easy to implement and widely supported by web technologies.
    3. HTTPS ensures secure communication through SSL/TLS encryption.

3. **Applications**: HTTP/HTTPS is commonly used in web-based IoT applications, such as smart home control systems and cloud-based analytics platforms.

6.3.2.2 WebSockets

1. **Overview**: WebSockets enable real-time, bidirectional communication between clients and servers over a single, long-lived connection. This protocol is particularly useful for applications that require low latency and real-time updates.
2. **Characteristics**:
    1. Reduces overhead compared to traditional HTTP requests by maintaining a persistent connection.
    2. Supports real-time messaging, making it ideal for applications with frequent data updates.
3. **Applications**: WebSockets are used in live data streaming applications, such as financial market data, gaming, and collaborative tools.

6.3.2.3 XMPP (Extensible Messaging and Presence Protocol)

1. **Overview**: XMPP is a communication protocol based on XML (Extensible Markup Language) designed for real-time messaging and presence information.
2. **Characteristics**:
    1. Supports extensibility, allowing developers to create custom features and applications.
    2. Provides real-time communication and presence information.
    3. Offers built-in security features, including encryption.
3. **Applications**: XMPP is commonly used in messaging applications, social media, and collaborative platforms.

6.4 Choosing the Right Communication Protocol

Selecting the appropriate communication protocol for an IoT application involves considering several factors:

6.4.1 Application Requirements

Understanding the specific requirements of the application is crucial. Factors such as data transmission frequency, latency tolerance, bandwidth constraints, and scalability needs will influence the choice of protocol.

6.4.2 Device Capabilities

The capabilities of the devices involved also play a significant role in protocol selection. Resource-constrained devices may require lightweight protocols like MQTT or CoAP, while more capable devices may support robust protocols like AMQP or HTTP.

6.4.3 Network Environment

The network environment in which the IoT system operates should be considered. For example, if the application relies on unreliable or low-bandwidth networks, protocols with built-in reliability features (such as MQTT) may be preferred.

6.4.4 Security Considerations

Security is paramount in IoT applications. Protocols that offer strong security mechanisms, such as encryption and authentication, should be prioritized, especially for applications handling sensitive data.

6.5 Challenges in IoT Communication Protocols

Despite their importance, several challenges are associated with IoT communication protocols:

6.5.1 Fragmentation of Protocols

The proliferation of various communication protocols can lead to fragmentation within the IoT ecosystem. This fragmentation makes interoperability between devices more challenging, as not all devices will support the same protocols.

6.5.2 Scalability Issues

As the number of connected devices continues to grow, scalability becomes a significant concern. Protocols must be designed to handle increasing data volumes and device connections without sacrificing performance.

6.5.3 Security Vulnerabilities

Security vulnerabilities in communication protocols can expose IoT devices to cyber threats. Implementing robust security measures, including encryption and authentication, is essential to mitigate these risks.

6.5.4 Power Consumption

For battery-operated IoT devices, power consumption is a critical factor. Some communication protocols may consume more power than others, affecting the longevity of devices. Choosing energy-efficient protocols is vital for extending battery life.

6.6 Future Trends in IoT Communication Protocols

As IoT technology continues to evolve, several trends are shaping the future of communication protocols:

6.6.1 Emergence of New Protocols

New communication protocols are being developed to address the specific needs of emerging IoT applications. These protocols focus on enhancing security, scalability, and efficiency, enabling more sophisticated IoT solutions.

6.6.2 Integration of AI and Machine Learning

The integration of artificial intelligence (AI) and machine learning into communication protocols can improve data transmission efficiency and optimize network performance. AI algorithms can analyze network conditions and adjust communication strategies accordingly.

6.6.3 Standardization Efforts

Efforts toward standardizing IoT communication protocols are gaining momentum. Standardization can facilitate interoperability, reduce fragmentation, and simplify the development of IoT solutions.

6.6.4 Enhanced Security Mechanisms

As cyber threats continue to evolve, communication protocols will need to incorporate advanced security mechanisms. This may include enhanced encryption methods, intrusion detection systems, and secure authentication protocols.

6.7 Conclusion

Communication protocols are a fundamental aspect of the Internet of Things, enabling devices to connect, communicate, and share data effectively. Understanding the various protocols available, their characteristics, and their applications is essential for designing robust IoT systems.

By carefully selecting the appropriate communication protocol based on application requirements, device capabilities, network environments, and security considerations, organizations can optimize their IoT implementations. Addressing the challenges associated with protocol fragmentation, scalability, and security will be critical to realizing the full potential of IoT.

As technology continues to advance, the future of IoT communication protocols will be shaped by emerging trends, including the development of new protocols, integration of AI, standardization efforts, and enhanced security mechanisms. Embracing these trends will enable organizations to create innovative IoT solutions that drive efficiency, enhance user experiences, and unlock new opportunities in an increasingly connected world.

Chapter 7: Security Challenges in IoT

7.1 Introduction to IoT Security

As the Internet of Things (IoT) continues to expand and integrate into various aspects of daily life, from smart homes to critical infrastructure, the need for robust security measures has become paramount. IoT devices collect, transmit, and store sensitive data, making them attractive targets for cybercriminals. Security vulnerabilities in IoT can lead to unauthorized access, data breaches, and potential disruptions in services. This chapter delves into the key security challenges facing IoT, explores the implications of these vulnerabilities, and discusses strategies for enhancing security in IoT systems.

7.2 Understanding IoT Security Challenges

7.2.1 Device Security Vulnerabilities

One of the primary challenges in IoT security is the inherent vulnerabilities present in devices. Many IoT devices are manufactured with limited processing power, making it difficult to implement robust security features. Common device vulnerabilities include:

1. **Weak Authentication Mechanisms**: Many IoT devices rely on default passwords or simplistic authentication methods, making them easy targets for unauthorized access. Weak authentication can allow attackers to take control of devices and misuse them for malicious purposes.
2. **Lack of Software Updates**: Some IoT devices are not designed to receive regular software updates, leaving them susceptible to known

vulnerabilities. Failure to update firmware can result in security holes that attackers can exploit.

3. **Insecure Communication Channels**: IoT devices often transmit data over insecure channels, which can be intercepted by attackers. Without proper encryption, sensitive data can be exposed during transmission.

7.2.2 Network Security Threats

The interconnected nature of IoT devices creates a complex network of communication, which can introduce several network security threats, including:

1. **DDoS Attacks**: Distributed Denial of Service (DDoS) attacks target IoT devices to overwhelm them with traffic, rendering them inoperable. Such attacks can disrupt services and lead to significant financial losses for organizations.
2. **Man-in-the-Middle (MitM) Attacks**: In MitM attacks, an attacker intercepts and alters communication between two devices. This can result in data manipulation or unauthorized access to sensitive information.
3. **Botnets**: Compromised IoT devices can be leveraged to create botnets—networks of infected devices controlled by attackers. Botnets can be used for various malicious activities, including DDoS attacks and credential stuffing.

7.2.3 Data Privacy Concerns

IoT devices collect vast amounts of data, raising significant privacy concerns. Security challenges related to data privacy include:

1. **Data Breaches**: Unauthorized access to IoT devices can lead to data breaches, exposing sensitive personal information. Such breaches can

have severe implications for individuals and organizations, including identity theft and reputational damage.

2. **Lack of User Awareness**: Many users are unaware of the data being collected by IoT devices and how it is used. This lack of transparency can lead to privacy violations and a distrust of IoT technologies.

3. **Data Misuse**: Collected data can be misused for malicious purposes, including targeted advertising and profiling without the user's consent. Organizations must ensure ethical data practices to protect user privacy.

7.3 Implications of Security Challenges

The security challenges faced by IoT have far-reaching implications for individuals, organizations, and society as a whole. Some of these implications include:

7.3.1 Economic Impact

Security breaches in IoT can lead to significant financial losses for organizations. The costs associated with data breaches, remediation efforts, legal liabilities, and reputational damage can be substantial. Additionally, the disruption of services can result in lost revenue and diminished customer trust.

7.3.2 Safety Risks

In critical infrastructure applications, such as healthcare, transportation, and energy, security vulnerabilities can pose serious safety risks. For example, unauthorized access to medical devices could endanger patient safety, while cyberattacks on transportation systems could disrupt operations and jeopardize public safety.

7.3.3 Erosion of Trust

As security incidents in IoT become more prevalent, public trust in IoT technologies may erode. Users may become hesitant to adopt IoT solutions if they perceive them as insecure or vulnerable to attacks. This erosion of trust can hinder the growth and development of IoT applications.

7.4 Strategies for Enhancing IoT Security

Addressing the security challenges in IoT requires a multi-faceted approach that encompasses various strategies and best practices:

7.4.1 Secure Device Design

1. **Implement Strong Authentication**: Manufacturers should prioritize strong authentication mechanisms, including multi-factor authentication (MFA), to protect devices from unauthorized access.
2. **Regular Software Updates**: IoT devices should be designed to receive regular firmware updates to patch vulnerabilities and enhance security. Manufacturers should provide clear mechanisms for users to update their devices easily.
3. **Encryption of Data**: Data transmitted between IoT devices should be encrypted using strong encryption algorithms to prevent interception and unauthorized access.

7.4.2 Network Security Measures

1. **Network Segmentation**: Organizations should segment their networks to isolate IoT devices from critical systems. This can help prevent attackers from gaining access to sensitive data and resources.
2. **Intrusion Detection and Prevention Systems (IDPS)**: Implementing IDPS can help monitor network traffic for suspicious activities and automatically respond to potential threats.

3. **Firewalls and Access Controls**: Firewalls and access control measures should be employed to restrict unauthorized access to IoT devices and networks. Role-based access control (RBAC) can ensure that users have appropriate permissions based on their roles.

7.4.3 User Education and Awareness

1. **Promote User Awareness**: Users should be educated about the security features of their IoT devices and the importance of strong passwords and regular updates. Awareness campaigns can help users understand potential risks and take proactive measures.
2. **Transparent Data Practices**: Organizations should communicate their data collection and usage practices transparently, ensuring that users are informed about how their data is used and stored.

7.4.4 Regulatory Compliance

1. **Adhere to Security Standards**: Organizations should comply with industry standards and regulations related to IoT security, such as the General Data Protection Regulation (GDPR) and the National Institute of Standards and Technology (NIST) guidelines. Compliance can help organizations establish a security framework and demonstrate commitment to protecting user data.
2. **Engage in Risk Assessments**: Regular risk assessments should be conducted to identify potential vulnerabilities in IoT systems and develop strategies to mitigate them. Organizations can adapt their security measures based on emerging threats and changing technologies.

7.5 Future Trends in IoT Security

The landscape of IoT security is continuously evolving, influenced by emerging technologies and changing threat vectors. Some future trends to consider include:

7.5.1 Artificial Intelligence and Machine Learning

AI and machine learning technologies can play a pivotal role in enhancing IoT security. These technologies can analyze vast amounts of data to identify patterns and detect anomalies that may indicate security threats. Predictive analytics can enable proactive measures to prevent attacks before they occur.

7.5.2 Zero Trust Architecture

The zero trust security model assumes that threats can exist both outside and inside the network. This approach emphasizes strict identity verification and continuous monitoring of devices and users. Implementing a zero trust architecture can enhance security in IoT environments by minimizing the risk of unauthorized access.

7.5.3 Blockchain Technology

Blockchain technology offers potential solutions for enhancing IoT security by providing decentralized and tamper-proof records of transactions and data exchanges. This can enhance data integrity, authentication, and transparency in IoT systems.

7.5.4 Standardization of Security Protocols

As the IoT ecosystem matures, efforts toward standardizing security protocols will likely increase. Standardization can promote interoperability, reduce fragmentation, and simplify the implementation of security measures across diverse devices and platforms.

7.6 Conclusion

Security challenges in the Internet of Things present significant risks to individuals, organizations, and society as a whole. Understanding these challenges and their implications is crucial for developing effective strategies to protect IoT systems from cyber threats.

By implementing robust security measures, including secure device design, network security protocols, user education, and regulatory compliance, organizations can enhance the security posture of their IoT implementations. The evolving landscape of IoT security necessitates a proactive approach, incorporating emerging technologies and best practices to address new threats.

As IoT continues to expand and integrate into various domains, ensuring the security of devices and data will be essential for fostering trust, protecting privacy, and realizing the full potential of IoT technology.

Chapter 8: Future Trends in IoT

8.1 Introduction

The Internet of Things (IoT) has transformed how we interact with technology, bridging the gap between the physical and digital worlds. As IoT continues to evolve, several trends are shaping its future, driving innovation and creating new opportunities across various sectors. This chapter explores the emerging trends in IoT, examining their implications for businesses, consumers, and society. By understanding these trends, stakeholders can prepare for the future landscape of IoT and leverage its potential to enhance productivity, improve quality of life, and address global challenges.

8.2 The Rise of Edge Computing

8.2.1 Definition and Importance

Edge computing refers to processing data closer to the source rather than relying on a centralized cloud server. This decentralized approach reduces latency, enhances data processing speeds, and alleviates bandwidth constraints, making it particularly beneficial for IoT applications.

8.2.2 Benefits of Edge Computing in IoT

1. **Reduced Latency**: By processing data locally, edge computing minimizes the delay in data transmission, allowing for real-time decision-making, which is crucial in applications such as autonomous vehicles and industrial automation.
2. **Bandwidth Efficiency**: Edge computing reduces the volume of data sent to the cloud, conserving bandwidth and reducing operational costs. Only essential data is transmitted, enabling more efficient use of network resources.

3. **Enhanced Privacy and Security**: Processing data at the edge can improve security and privacy by minimizing the transmission of sensitive information over networks. Local data processing can mitigate risks associated with data breaches during transmission.

8.2.3 Use Cases

Edge computing is becoming increasingly relevant in various IoT applications, including:

1. **Smart Cities**: In smart city projects, edge computing can analyze data from sensors in real time to optimize traffic flow, enhance public safety, and improve resource management.
2. **Healthcare**: Wearable health devices can process data locally to provide immediate feedback to users and healthcare providers, ensuring timely interventions.
3. **Manufacturing**: In industrial settings, edge computing enables real-time monitoring of machinery, predictive maintenance, and improved operational efficiency by analyzing data on-site.

8.3 The Integration of Artificial Intelligence (AI)

8.3.1 AI and IoT Synergy

The integration of artificial intelligence (AI) with IoT is a significant trend that enhances the capabilities of connected devices. AI algorithms can analyze vast amounts of data generated by IoT devices, uncovering patterns, predicting outcomes, and enabling automated decision-making.

8.3.2 Benefits of AI in IoT

1. **Predictive Analytics**: AI can analyze historical data to predict future trends, enabling proactive decision-making in various sectors, from

manufacturing to healthcare. For example, predictive maintenance can anticipate equipment failures before they occur.

2. **Smart Automation**: AI-powered IoT devices can learn from user behavior and environmental changes to automate processes, improving efficiency and user experience. Smart home systems can adjust heating and lighting based on occupants' preferences and habits.

3. **Enhanced Decision-Making**: AI can provide actionable insights derived from real-time data analysis, empowering organizations to make data-driven decisions that enhance operational efficiency and customer satisfaction.

8.3.3 Use Cases

AI integration in IoT is being realized in numerous applications:

1. **Smart Homes**: AI algorithms in smart home devices can analyze user patterns to optimize energy consumption, enhance security, and improve overall comfort.

2. **Healthcare**: AI-powered IoT devices can monitor patients' health in real time, providing alerts and recommendations based on their conditions and historical data.

3. **Agriculture**: AI-enabled IoT devices can analyze soil conditions, weather patterns, and crop health, optimizing irrigation and fertilization processes for improved yields.

8.4 The Expansion of 5G Technology

8.4.1 Overview of 5G Technology

5G technology is the fifth generation of mobile networks, offering significantly faster data speeds, lower latency, and increased capacity compared to its predecessors. This technological advancement is set to revolutionize the IoT landscape by enabling more devices to connect seamlessly and transmit data in real-time.

8.4.2 Impact of 5G on IoT

1. **Increased Device Connectivity**: 5G can support a higher density of connected devices, allowing for the proliferation of IoT applications across urban and rural areas.
2. **Real-Time Data Transmission**: The low latency of 5G enables real-time data exchange, making it ideal for applications requiring immediate responses, such as autonomous vehicles and remote surgery.
3. **Enhanced Reliability**: 5G networks offer improved reliability and consistency in connectivity, ensuring that critical IoT applications remain functional even in high-demand scenarios.

8.4.3 Use Cases

The advent of 5G technology is paving the way for innovative IoT applications:

1. **Autonomous Vehicles**: The low latency and high bandwidth of 5G are essential for the safe operation of autonomous vehicles, enabling them to communicate with each other and their surroundings in real time.
2. **Smart Cities**: 5G will enhance smart city initiatives by supporting a vast number of connected sensors and devices, facilitating better traffic management, environmental monitoring, and public safety.
3. **Telemedicine**: 5G technology enables remote healthcare services by supporting high-definition video consultations and real-time monitoring of patients' health data, even in rural areas.

8.5 Sustainability and IoT

8.5.1 The Role of IoT in Promoting Sustainability

As global awareness of environmental issues grows, IoT is emerging as a powerful tool for promoting sustainability. IoT applications can help organizations and individuals monitor and reduce their environmental impact.

8.5.2 Benefits of IoT for Sustainability

1. **Energy Management**: IoT devices can optimize energy consumption by monitoring usage patterns and adjusting settings accordingly. Smart meters and smart grids enable real-time energy management, reducing waste and costs.
2. **Resource Conservation**: IoT solutions can monitor water usage, air quality, and waste management, providing insights that help organizations and communities conserve resources and minimize pollution.
3. **Supply Chain Optimization**: IoT can enhance supply chain transparency and efficiency, allowing organizations to track products in real time and reduce waste through better inventory management.

8.5.3 Use Cases

IoT's role in promoting sustainability is evident in various applications:

1. **Smart Agriculture**: IoT devices can monitor soil conditions, weather, and crop health, enabling farmers to optimize resource use and reduce environmental impact.
2. **Waste Management**: IoT sensors can monitor waste levels in bins, optimizing collection routes and schedules to reduce fuel consumption and improve efficiency.

3. **Building Management**: Smart building systems can monitor energy usage and optimize heating, ventilation, and air conditioning (HVAC) systems to minimize energy consumption and improve comfort.

8.6 Enhanced Security Measures

8.6.1 The Growing Focus on Security

As IoT adoption increases, so do concerns about security vulnerabilities. Organizations are recognizing the need for enhanced security measures to protect devices, data, and user privacy.

8.6.2 Trends in IoT Security

1. **Zero Trust Security Models**: Organizations are adopting zero trust security principles, which require continuous verification of users and devices, regardless of their location. This approach minimizes the risk of unauthorized access and data breaches.
2. **Advanced Encryption Techniques**: The use of advanced encryption methods is becoming standard practice in securing data transmitted between IoT devices. Strong encryption helps protect sensitive information from interception.
3. **Regulatory Compliance**: Governments and regulatory bodies are implementing stricter regulations governing IoT security and data protection. Compliance with these regulations is becoming essential for organizations to mitigate risks and protect consumer trust.

8.6.3 Use Cases

Enhanced security measures are critical in various IoT applications:

- **Healthcare**: Protecting sensitive patient data transmitted by IoT medical devices is crucial for maintaining patient confidentiality and regulatory compliance.

- **Smart Cities**: Ensuring the security of connected infrastructure, such as traffic management systems and public safety devices, is vital to preventing disruptions and protecting citizens.
- **Industrial IoT**: Securing industrial IoT systems is essential to protect against cyber threats that could disrupt operations and compromise safety.

8.7 Conclusion

The future of the Internet of Things is characterized by rapid technological advancements and evolving trends that promise to reshape industries and enhance quality of life. As edge computing, AI integration, 5G technology, sustainability initiatives, and enhanced security measures continue to gain traction, stakeholders must remain vigilant and adaptable.

By embracing these trends, organizations can unlock new opportunities, optimize processes, and contribute to a more connected and sustainable world. As we move forward, collaboration among technology providers, policymakers, and consumers will be essential to navigate the challenges and seize the potential of IoT.

In summary, the future of IoT holds immense promise, and those who harness its capabilities will be well-positioned to thrive in an increasingly connected landscape. By staying informed about emerging trends and proactively addressing security challenges, we can create a safer, more efficient, and sustainable future powered by the Internet of Things.

Chapter 9: Real-World Applications of IoT

9.1 Introduction

The Internet of Things (IoT) is revolutionizing various industries by enabling connected devices to communicate, share data, and perform tasks autonomously. From enhancing operational efficiency to improving user experiences, IoT applications are transforming how businesses and individuals interact with technology. This chapter explores the diverse real-world applications of IoT across different sectors, highlighting specific use cases that illustrate the profound impact of IoT technology.

9.2 Smart Homes

9.2.1 Overview

Smart home technology is one of the most recognizable applications of IoT, providing homeowners with enhanced control over their living environments. By connecting devices such as thermostats, lights, and security systems to the internet, smart homes offer convenience, energy efficiency, and improved security.

9.2.2 Key Use Cases

1. **Smart Thermostats**: Devices like the Nest Learning Thermostat automatically adjust heating and cooling based on user preferences and behaviors, leading to energy savings and enhanced comfort. These devices can be controlled remotely through smartphones, allowing users to optimize energy usage even when away from home.

2. **Smart Lighting**: IoT-enabled lighting systems, such as Philips Hue, allow users to control the brightness and color of their lights through mobile apps or voice commands. This not only enhances convenience but also promotes energy efficiency by enabling users to set schedules and automate lighting based on occupancy.

3. **Home Security Systems**: Smart security cameras, doorbells, and locks enable homeowners to monitor their properties in real time. With features like motion detection and remote access, users can receive alerts and view live feeds from their smartphones, enhancing home security and peace of mind.

9.2.3 Benefits

The integration of IoT in smart homes leads to increased energy efficiency, enhanced security, and improved quality of life. By automating routine tasks and providing remote access to home systems, smart homes contribute to a more comfortable and secure living environment.

9.3 Healthcare

9.3.1 Overview

IoT is making significant strides in the healthcare sector, transforming patient monitoring, diagnosis, and treatment. IoT devices enable healthcare providers to collect real-time data, improve patient outcomes, and enhance the overall quality of care.

9.3.2 Key Use Cases

1. **Wearable Health Devices**: Devices such as Fitbit and Apple Watch monitor vital signs, activity levels, and sleep patterns. By collecting and analyzing this data, healthcare providers can offer personalized recommendations and early interventions based on individual health trends.

2. **Remote Patient Monitoring**: IoT devices enable healthcare professionals to monitor patients with chronic conditions from home. For instance, devices that measure blood glucose levels in diabetic patients can automatically transmit data to healthcare providers, allowing for timely adjustments in treatment plans.
3. **Smart Pill Bottles**: IoT-enabled pill bottles remind patients to take their medications on time and notify healthcare providers if doses are missed. This technology enhances medication adherence and improves health outcomes for patients with chronic conditions.

9.3.3 Benefits

The adoption of IoT in healthcare leads to improved patient outcomes, enhanced efficiency in care delivery, and reduced healthcare costs. By enabling continuous monitoring and data analysis, IoT empowers patients and providers to make informed decisions about health and wellness.

9.4 Industrial IoT (IIoT)

9.4.1 Overview

Industrial IoT (IIoT) refers to the use of IoT technology in manufacturing and industrial applications. IIoT enables organizations to optimize operations, improve safety, and increase productivity through data-driven insights and automation.

9.4.2 Key Use Cases

1. **Predictive Maintenance**: IIoT devices monitor the condition of machinery in real-time, allowing organizations to predict equipment failures before they occur. By analyzing data from sensors, businesses can schedule maintenance proactively, reducing downtime and minimizing repair costs.

2. **Supply Chain Optimization**: IoT devices track inventory levels, shipments, and equipment usage throughout the supply chain. This visibility enables organizations to optimize logistics, reduce waste, and improve overall efficiency in their operations.

3. **Worker Safety**: IoT-enabled wearables, such as smart helmets and safety vests, monitor workers' vital signs and environmental conditions in hazardous work environments. These devices can alert supervisors to potential safety risks, enhancing employee safety and compliance with regulations.

9.4.3 Benefits

The implementation of IIoT results in improved operational efficiency, reduced costs, and enhanced safety in industrial environments. By leveraging data analytics and automation, organizations can make informed decisions that drive productivity and innovation.

9.5 Agriculture

9.5.1 Overview

IoT is transforming agriculture by enabling precision farming and enhancing resource management. By connecting sensors and devices in the field, farmers can optimize crop yields, conserve resources, and improve overall sustainability.

9.5.2 Key Use Cases

1. **Soil Monitoring**: IoT sensors monitor soil moisture, temperature, and nutrient levels, providing farmers with real-time data to optimize irrigation and fertilization. This precision farming approach leads to improved crop health and reduced resource waste.

2. **Livestock Tracking**: IoT devices are used to monitor the health and location of livestock. Wearable sensors can track vital signs, activity

levels, and grazing patterns, allowing farmers to identify health issues early and optimize grazing strategies.

3. **Weather Monitoring**: IoT-enabled weather stations collect real-time data on local weather conditions, enabling farmers to make informed decisions about planting, harvesting, and resource management.

9.5.3 Benefits

The integration of IoT in agriculture leads to increased efficiency, reduced resource consumption, and improved crop yields. By leveraging data-driven insights, farmers can enhance productivity and sustainability in their operations.

9.6 Transportation and Logistics

9.6.1 Overview

IoT is revolutionizing the transportation and logistics industry by enabling real-time tracking, route optimization, and improved fleet management. By connecting vehicles and infrastructure, organizations can enhance operational efficiency and customer satisfaction.

9.6.2 Key Use Cases

1. **Fleet Management**: IoT devices track the location, speed, and condition of vehicles in real time. This data allows logistics companies to optimize routes, reduce fuel consumption, and improve delivery times.
2. **Asset Tracking**: IoT sensors are used to monitor the location and condition of shipments throughout the supply chain. Real-time tracking provides visibility into the movement of goods, enabling companies to respond quickly to delays or issues.
3. **Smart Traffic Management**: IoT technology enables cities to monitor traffic patterns and adjust signals in real time. This can reduce

congestion, improve public transportation efficiency, and enhance overall traffic flow.

9.6.3 Benefits

The application of IoT in transportation and logistics leads to enhanced operational efficiency, reduced costs, and improved customer satisfaction. By leveraging real-time data, organizations can make informed decisions that optimize their supply chain and transportation processes.

9.7 Retail

9.7.1 Overview

IoT is transforming the retail industry by enhancing the shopping experience, optimizing inventory management, and improving customer engagement. Retailers are increasingly leveraging IoT technology to gain insights into consumer behavior and streamline operations.

9.7.2 Key Use Cases

1. **Smart Shelves**: IoT-enabled shelves can monitor inventory levels in real time, notifying retailers when stock is running low. This automation improves inventory management and reduces the likelihood of stockouts.
2. **Personalized Shopping Experiences**: Retailers can use IoT devices to collect data on customer preferences and behaviors. This data allows for personalized marketing campaigns and targeted promotions, enhancing customer engagement and loyalty.
3. **Smart Checkout Systems**: IoT technology enables frictionless checkout experiences through the use of mobile apps and smart carts. Customers can scan items as they shop and pay seamlessly, reducing wait times and improving the overall shopping experience.

9.7.3 Benefits

The integration of IoT in retail leads to improved inventory management, enhanced customer experiences, and increased sales. By leveraging data-driven insights, retailers can optimize their operations and better meet customer needs.

9.8 Conclusion

The real-world applications of IoT span diverse industries, showcasing its transformative potential across various sectors. From smart homes to healthcare, agriculture, and logistics, IoT is driving innovation, enhancing efficiency, and improving quality of life.

As organizations continue to explore and implement IoT solutions, the focus will shift toward maximizing the benefits of connected devices while addressing challenges such as security, data privacy, and interoperability. By embracing IoT technology, businesses can position themselves for success in an increasingly connected world, unlocking new opportunities and driving growth in their respective industries.

Chapter 10: Challenges and Risks in IoT

10.1 Introduction

The Internet of Things (IoT) holds immense potential to transform industries, enhance daily life, and drive economic growth. However, the rapid proliferation of IoT devices and applications brings forth a myriad of challenges and risks that must be addressed to fully realize its benefits. This chapter delves into the critical challenges facing IoT, including security vulnerabilities, data privacy concerns, interoperability issues, scalability, and regulatory compliance. Understanding these challenges is essential for stakeholders—businesses, consumers, and policymakers— to mitigate risks and create a secure, efficient, and sustainable IoT ecosystem.

10.2 Security Vulnerabilities

10.2.1 Overview

Security is one of the most pressing challenges in IoT. With billions of connected devices transmitting sensitive data, the risk of cyberattacks, data breaches, and unauthorized access is heightened. Many IoT devices are designed with convenience in mind, often at the expense of robust security measures.

10.2.2 Common Security Issues

1. **Weak Authentication**: Many IoT devices use default passwords or weak authentication methods, making them susceptible to unauthorized access. Attackers can exploit these vulnerabilities to gain control of devices and networks.

2. **Insecure Communication**: Many IoT devices do not encrypt data during transmission, making it easier for hackers to intercept and manipulate data. Without secure communication protocols, sensitive information is at risk.
3. **Lack of Software Updates**: IoT devices often lack mechanisms for timely software updates and patches. This creates security gaps, as vulnerabilities remain unaddressed, leaving devices exposed to attacks.

10.2.3 Potential Consequences

The consequences of security vulnerabilities in IoT can be severe, including:

1. **Data Breaches**: Unauthorized access to sensitive information can lead to significant data breaches, resulting in financial losses and reputational damage for organizations.
2. **Service Disruptions**: Cyberattacks targeting IoT systems can disrupt critical services, such as healthcare, transportation, and utilities, endangering public safety and well-being.
3. **Physical Damage**: In cases where IoT devices control physical systems (e.g., industrial machinery or smart grids), cyberattacks can result in physical damage, posing risks to human safety and infrastructure.

10.3 Data Privacy Concerns

10.3.1 Overview

As IoT devices collect vast amounts of personal data, data privacy concerns have become increasingly prominent. Consumers are often unaware of how their data is collected, stored, and used, raising questions about consent and transparency.

10.3.2 Key Privacy Issues

1. **Data Ownership**: The question of who owns the data generated by IoT devices is complex. Users may feel that they should retain ownership of their data, while manufacturers and service providers may assert rights over the data generated by their devices.
2. **Informed Consent**: Many users do not fully understand the implications of sharing their data with IoT devices. Inadequate transparency and complex privacy policies can prevent users from making informed choices about their data.
3. **Data Misuse**: There is a risk that personal data collected by IoT devices could be misused for targeted advertising, surveillance, or other purposes without user consent.

10.3.3 Potential Consequences

Data privacy concerns can have far-reaching consequences, including:

1. **Loss of Trust**: If consumers perceive that their privacy is compromised, they may lose trust in IoT technology, hindering adoption and innovation in the sector.
2. **Regulatory Backlash**: Governments may implement stricter regulations to protect consumer privacy, leading to compliance challenges for organizations that rely on IoT technology.
3. **Legal Liability**: Companies that fail to adequately protect consumer data may face legal actions and financial penalties, impacting their bottom line and reputation.

10.4 Interoperability Issues

10.4.1 Overview

Interoperability refers to the ability of different IoT devices and systems to work together seamlessly. Achieving interoperability is a significant challenge due to the lack of standardized protocols and frameworks.

10.4.2 Challenges to Interoperability

1. **Diverse Standards**: The IoT ecosystem comprises a wide range of devices from various manufacturers, each using different communication protocols and standards. This diversity complicates integration and communication between devices.
2. **Vendor Lock-In**: Organizations may become reliant on specific vendors for IoT solutions, limiting their ability to integrate devices from other manufacturers. This can lead to increased costs and reduced flexibility.
3. **Fragmented Ecosystem**: The lack of a unified IoT ecosystem can hinder the development of innovative applications and services that rely on data from multiple sources. Fragmentation can also create silos of information that impede data sharing and collaboration.

10.4.3 Potential Consequences

Interoperability issues can result in:

1. **Increased Costs**: Organizations may face higher costs associated with integrating disparate systems, leading to inefficient operations and reduced return on investment.
2. **Limited Functionality**: The inability to integrate devices can limit the functionality of IoT applications, preventing users from fully realizing the benefits of connected technology.
3. **Slower Adoption**: Interoperability challenges can deter businesses from adopting IoT solutions, slowing down innovation and hindering the growth of the IoT market.

10.5 Scalability Challenges

10.5.1 Overview

As the number of connected devices continues to grow, scalability becomes a critical challenge for IoT systems. Organizations must ensure that their infrastructure can accommodate increased device counts and data volumes without compromising performance.

10.5.2 Key Scalability Issues

4. **Infrastructure Limitations**: Many existing IoT systems are built on legacy infrastructure that may not be capable of supporting the scale of new IoT deployments. Upgrading infrastructure can be costly and time-consuming.
5. **Data Management**: The exponential growth of data generated by IoT devices can overwhelm traditional data management systems. Organizations must invest in scalable data storage and processing solutions to handle large volumes of data effectively.
6. **Network Congestion**: As more devices connect to the internet, network congestion can become a significant issue, leading to latency, reduced performance, and unreliable connections.

10.5.3 Potential Consequences

Scalability challenges can lead to:

4. **Performance Issues**: Organizations may experience degraded performance as systems struggle to accommodate increased device loads, leading to delays and disruptions in service.
5. **Increased Operational Costs**: The need to upgrade infrastructure and invest in data management solutions can result in increased operational costs, impacting the overall profitability of IoT initiatives.
6. **Limited Growth Opportunities**: Organizations that cannot scale their IoT systems effectively may miss out on growth opportunities and fail to keep pace with competitors.

10.6 Regulatory Compliance

10.6.1 Overview

The regulatory landscape for IoT is evolving, with governments and regulatory bodies implementing guidelines and standards to protect consumers and ensure data security. Compliance with these regulations can be challenging for organizations operating in the IoT space.

10.6.2 Key Compliance Challenges

1. **Varying Regulations**: Different regions may have varying regulations governing data privacy, security, and consumer protection. Organizations operating globally must navigate a complex landscape of compliance requirements.
2. **Evolving Standards**: As IoT technology continues to evolve, regulations and standards may change, requiring organizations to adapt their practices to remain compliant. Staying abreast of these changes can be resource-intensive.
3. **Lack of Clarity**: In some cases, regulations may lack clarity, leaving organizations uncertain about their obligations and potential liabilities. This uncertainty can hinder decision-making and strategic planning.

10.6.3 Potential Consequences

Regulatory compliance challenges can result in:

- **Legal Risks**: Non-compliance with regulations can lead to legal actions, fines, and penalties, negatively impacting an organization's financial health and reputation.

- **Increased Costs**: Organizations may incur additional costs associated with compliance efforts, such as implementing data protection measures and hiring legal experts.
- **Reputation Damage**: Failing to comply with regulations can damage an organization's reputation, eroding consumer trust and hindering future growth.

10.7 Conclusion

The challenges and risks associated with IoT are significant and multifaceted. As the technology continues to advance and permeate various sectors, stakeholders must proactively address these challenges to mitigate risks and maximize the benefits of IoT.

By prioritizing security, ensuring data privacy, fostering interoperability, addressing scalability, and navigating regulatory compliance, organizations can create a robust and resilient IoT ecosystem. Collaborative efforts among technology providers, policymakers, and consumers will be essential in overcoming these challenges and realizing the full potential of the Internet of Things.

As IoT technology evolves, continuous innovation and adaptation will be crucial for building a secure and sustainable future, ensuring that the advantages of IoT are accessible to all while safeguarding against the inherent risks and challenges.